CLASSIC STEAM TRAINS

STEAM SURVIVORS AROUND THE WORLD

CLASSIC STEAM TRAINS

STEAM SURVIVORS AROUND THE WORLD

COLIN GARRATT

LORENZ BOOKS

This edition is published by Lorenz Books

Lorenz Books is an imprint of Anness Publishing Ltd
Hermes House, 88–89 Blackfriars Road, London SE1 8HA
tel. 020 7401 2077; fax 020 7633 9499
www.lorenzbooks.com; info@anness.com

This edition distributed in the UK by Aurum Press Ltd,
25 Bedford Avenue, London WC1B 3AT;
tel. 020 7637 3225; fax 020 7580 2469

This edition distributed in the USA and Canada by National Book Network,
4720 Boston Way, Lanham, MD 20706;
tel. 301 459 3366; fax 301 459 1705; www.nbnbooks.com

This edition distributed in Australia by Pan Macmillan Australia,
Level 18, St Martins Tower, 31 Market St, Sydney, NSW 2000;
tel. 1300 135 113; fax 1300 135 103; customer.service@macmillan.com.au

This edition distributed in New Zealand by David Bateman Ltd,
30 Tarndale Grove, Off Bush Road, Albany, Auckland;
tel. (09) 415 7664; fax (09) 415 8892

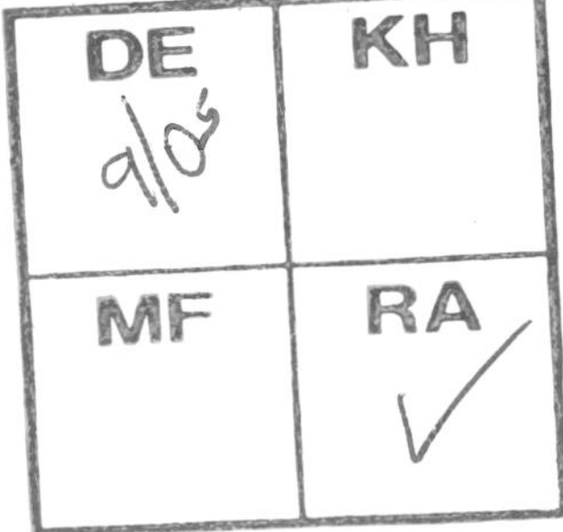

A CIP catalogue record for this book is available from the British Library.

Publisher: Joanna Lorenz
Project Editor: Charlotte Berman
Designers: Paul Hunt - Milepost 92½, Andrea Bettella – Artmedia
Production Controller: Joanna King
Editorial Reader: Hayley Kerr

1 3 5 7 9 10 8 6 4 2

CONTENTS

FOREWORD

THE END OF AN ERA

Sitting in the warm afternoon sunshine in 1999 on the grassy banks of India's last steam main line, between Wankaner and Morbi, I found it hard to believe that it was nearly all over. It was my fiftieth year of sitting on grassy banks watching steam trains and I wondered where on earth I would ever again encounter such joys. Neighbouring China offered some possibilities but even there – a country that had over 10,000 steam locomotives active in the mid-1980s – the declared intention is to be rid of steam on the national railway network by the autumn of the year 2000.

Restored steam railways, which attract many tourists every year, will keep the torch burning for steam, but the sight and sound of a real working steam train set against the great landscapes of the world is a thing of great beauty. The howling of a wolf on the Siberian Plains, or the distant roar of a lion on a dark jungle night are moments that cannot be reproduced in the confines of a zoo. "Captive" locomotives are equally anaesthetized.

Sitting on that bank in India, I thought of the hundreds of thousands of steam locomotives known to have been built over almost two centuries and how the vast majority had never been photographed in colour or, in many cases, at all. I reflected on my thirty-year commitment to document the last steam locomotives of

LEFT: *One of the world's last Texas type 2-10-4s on Brazil's Teresa Cristina Railway. These engines represent American steam super power scaled down for metre (3¼ ft) gauge operations. They have a grate area of 6.5 sq m (70 sq ft) and can haul 1,800 tonne (1,771 ton) coal trains at speeds of up to 96.5 kph (60 mph).*

ABOVE: *An Indian Railways standard YG Class metre (3¼ ft) gauge Mikado on the coaling road at Wankaner at the very end of main-line steam operation in India. The shed labourers empty the coal on to the ground from the broad-gauge wagons and manually coal the metre (3¼ ft) gauge engines with wicker baskets.*

OPPOSITE: *Sunlight, smoke and shadow give an eerie feel to a Chinese SY Class 2-8-2 residing amid the sooty magic of Anshan loco shed.*

the world and how, despite having travelled to some fifty countries, the pictures I have taken are but a blink amid an infinity of industrial legend.

That our lives are poorer for the passing of such treasures can hardly be doubted. The disappearance of the steam train has helped to produce an increasingly colourless and mundane world.

The decline in the significance of rail transport over the last fifty years has been a massively retrograde step and, of course, another factor in the disappearance of steam traction. Even if one were to accept the argument that steam was uneconomic, how can greysuits throughout the world justify spending countless billions of pounds on road transportation, a system which is both socially and environmentally inferior to rail?

Such thoughts were interrupted by the 16.50 from Morbi to Wankaner. As the distant rhythms became audible, a swirling sooty haze could be seen curling across the landscape like an approaching tornado. The rasping beat of the engine was irregular because of indifferent maintenance, while the poor track conditions caused the engine to lurch alarmingly from side to side. With a piercing screech, the engine sent a herd of goats which had been grazing at the lineside into a wild stampede. Though begrimed and work weary, the elegant lines of the YP 4-6-2 looked as magnificent as ever as the high-stepping Pacific swept past, issuing wafts of oily-smelling sulphur across the embankments. A cheery wave from the crew, the flash of a white-hot fire, a glimpse of the rich tapestry of Indian humanity as the coaches swept by – including a beggar standing in the open doorway of the rear vehicle, having plied his trade throughout the train – and it was gone.

Later that evening I went down to the loco shed at Wankaner. The Pacific had by then passed through the ash pit, been coaled and watered and stood quietly simmering in readiness for its next turn of duty. It was a scene that took me back in time: the dripping water column looked like a pillar from some ancient Greek temple;

ABOVE: A Chinese YJ Class industrial Prairie 2-6-2 waits to draw ladles of molten iron away from blast furnace No.2 at Anshan Iron and Steel Works.

RIGHT: The footplate crew during a night turn on the sugar plantations of Campos state, Brazil.

coal-blackened labourers shovelled ash out of the pits; pigeons cooed in the blackened rafters and the shed cat, tired of pursuing lizards, dozed outside the foreman fitter's office. Shrouds of acrid smoke coursed through the shed from a YG which had been lit up for a freight turn the following day, while fitters flitted from engine to engine, fighting a rearguard action to keep alive, against all odds, these remnants of a bygone age. I was reminded of what I have always known, that the steam locomotive means hard, back-breaking toil in conditions which are anti-social. These living remnants of the smoke-stack age were undeniably incongruous and looked like dinosaurs from a long-distant past.

An engine whistle resounded deep within the shed, followed by a roar of steam from the cylinder cocks. Steam filled the shed as the first heavy cough of exhaust was followed by a wheel slip, which sent a black cloud of smoke up through the roof and high into the azure evening sky. The shed momentarily disappeared in clouds of

TOP LEFT: A former Darjeeling Himalayan B Class 0-4-0ST enjoys a further lease of active life on the 0.6m (2ft) gauge colliery network at Tipong on the Assam coalfield.

ABOVE LEFT: Oil cans and grease guns belonging to "El Esla", a Sharp Stewart 0-6-0T of 1885, at Sabero colliery in northern Spain.

ABOVE: Blacksmith at the Location Locomotive Works, Ghana.

ABOVE RIGHT: An Italian State Railways 625 Class 2-6-0 on a local passenger train.

BELOW RIGHT: Italian State Railways 743 Class 2-8-0 No. 301, fitted with a Franco Crosti boiler, heads the 10.50 goods train from Pavia to Cremona through Corteolono on 24 April 1976.

FAR RIGHT: A Robert Stephenson and Hawthorn 1.7 m (5 ft 6 in) gauge 1944 0-4-0ST at Kulti Iron Works in Bengal.

ABOVE: One of the last surviving Indian Railways 1.7 m (5 ft 6 in) gauge WG Class 2-8-2s, No. 9428, ends its days in industrial service at Chunar Cement Works. The engine was built by the Canadian Locomotive Works in 1956.

LEFT: A hastily applied adornment to a Chinese industrial locomotive.

BELOW RIGHT: Number plate of China Railways' JS Class 2-8-2 No. 6004, one of a few members of this standard main-line type which have passed into industrial service.

steam until, slipping and wheezing, the rusty YG emerged, dragging two dead engines out into the yard. Slipping and sliding along the greasy rails, the engine sent tremors through the ground.

The camaraderie of the shed staff – the labourers, the steam raisers, boiler washers, footplate crews and foremen – was the result of responding to the challenge of taming these aged machines and keeping them running. Steam always provided a challenge which brought out the best in men in a way that modern traction never can. The raw elements of fire and water are the fundamentals of creation itself and when encapsulated in the shell of the steam locomotive – one of the most beautiful and sensuous creations of all time – both adult and child respond alike.

Leaving Wankaner that evening I thought of the curtain about to fall on one of the greatest legends of all time; that I have lived within that legend will be a constant and incomparable joy for the rest of my days.

Colin Garratt
Milepost 92½,
Newton Harcourt
Leicestershire
England

625 164

№ 27

INDUSTRIAL WORKHORSES

COALFIELDS

THE WORLD'S FIRST STEAM LOCOMOTIVES WERE INDUSTRIALS EMPLOYED IN BRITISH IRONWORKS AND COLLIERIES. AS RAILWAYS DEVELOPED BEYOND THE CONFINES OF INDUSTRY, LARGER LOCOMOTIVES QUICKLY EVOLVED WITH SEPARATE TYPES BEING DEVELOPED FOR HAULING GOODS AND PASSENGER TRAINS. INDUSTRIAL LOCOMOTIVES DEVELOPED AS A SEPARATE ENTITY; OFTEN HIDDEN AWAY AMID THE CONFINES OF LARGE FACTORIES, THEY WERE RESPONSIBLE FOR MOVING MATERIALS AROUND THE PLANTS AND CONNECTING WITH THE MAIN-LINE RAILWAY.

Steam locomotives were used on British coalfields for almost one-and-three-quarter centuries. As early as 1812 an example appeared in Yorkshire, and a few years later some of George Stephenson's early locomotives were put to work at the historic Northumberland coalfield. Coal and steam were the lifeblood of the Industrial Revolution. During the 19th century, hundreds of collieries were developed, most of which used steam traction to convey coal to the main-line railway or to waterways or docks, and as late as the early 1960s the National Coal Board were ordering new locomotives. It seemed that the use of steam would last indefinitely in Britain's coalfields. No one imagined that by the end of the century not only would steam have vanished but the vast majority of British collieries would have closed.

Most of the steam locomotives used at collieries throughout the world were four- or six- wheeled tanks, particularly in Britain and Europe, but larger locomotives were needed where the connection from the colliery to the main-line railway was lengthy or steeply graded. Coalfields also used redundant main-line engines; some of the best examples occurred in South Africa, where powerful 4-8-2s and Garratts were employed. Some of the last steam locomotives built were for coalfield service; China's SY Class 2-8-2s were still being delivered new as late as 1997. These Mikados were similar to the type of engine used on America's secondary main lines prior to World War I and their occurrence in colliery service provides a graphic illustration of how wagon sizes and train weights have increased since the early days of railways. A handful of SYs will almost certainly survive until 2012, rendering steam a continuous source of motive power for coalfields for two centuries.

TOP: This historic painting by David Weston depicts an early industrial locomotive in a British colliery.

LEFT: A vintage 0.6 m (2 ft) gauge 0-4-0 Saddle Tank, built by Bagnalls of Stafford for the Assam Railway & Trading Company, is worked at a hillside mine with a Lancashire boiler house in the background.

OPPOSITE: A typical British Industrial Saddle Tank at Pennyvenie mine in Ayrshire. The engine is one of a standard design introduced by Andrew Barclay of Kilmarnock.

ABOVE RIGHT: A typical British colliery scene at Desford, Leicestershire. Like so many of Britain's small collieries, this pit has closed and no trace remains at this once thriving location.

RIGHT: An Afrikaner driver on a Gauteng coalfield, South Africa.

N. C. B.
Nº 21
WEST AYR AREA
N C B

Industrial Workhorses

Sugar Plantations

Many of the world's most interesting steam survivors are to be found in sugar plantation service, notably in Cuba, Java, India and the Philippines. Cuba is the last bastion of classic American steam, while Java, once part of the Dutch East Indies, has an amazing variety of veterans from legendary German and Dutch builders. Steam survivors in India and the Philippines also reflect their country's colonial past with British and American locomotives respectively.

LEFT TOP: A Hunslet 0-4-2ST hauls cane on the Trangkil Sugar system in Java. This humble Saddle Tank left Hunslet's works in 1971 and was the last of tens of thousands of steam locomotives exported from Britain for service on railways all over the world.

LEFT BOTTOM: The cane yard at the sugar mill on the Philippine island of Negros. One of the company's Dragons, in the form of a 1924 Baldwin 0-6-2ST, acts as works shunter.

NEAR LEFT: An American-inspired wooden signal box with Baldwin Mogul in Cuba.

TOP: A splendid Orenstein and Koppel 0-10-0 tender/tank at Pagottan Mill, Java.

ABOVE: The builder's plate from the Hunslet 0-4-2ST pictured top left.

RIGHT: An Orenstein and Koppel 0-8-0 tender/tank with sandman at Meritjan Mill, Java. Sandmen sit on the engine's buffer beam with a huge tray of sand, which they spray manually on to the track whenever the rails are wet or muddy from the frequent tropical storms which lash Java during the milling season.

Industrial Workhorses

Iron & Steel

In 1803, Richard Trevithick, a Cornish mining engineer, built the world's first steam locomotive at Coalbrookdale Ironworks. It was not a success, but the following year Trevithick produced a locomotive to work on the Pennydarren Tramway in South Wales where the engine, taking the place of horses, drew a train of iron from the works to the canal basin at Merthyr, so beginning steam's auspicious relationship with ironworks service.

The iron and steel industry was one of the most demanding of the Industrial Revolution. A vast amount of raw materials had to be conveyed to, around and out of the complexes, and most establishments were reliant on railways. The dynamism of the complexes, with their smouldering coke ovens and blast furnaces, was matched by the locomotives as, spitting smoke, steam and fire, they threaded their way through the shadowy, grime-laden structures.

As a general rule, six- or eight-wheeled tank engines predominated, especially when clearances between fabrication sheds were restricted. Some complexes, such as Corby in the English Midlands, were located on the ironstone bed, and the network of railways ran deep into the surrounding countryside, bringing ore to the complex. Corby's locomotives were divided between the mines and the steel works, and operated from separate running sheds. A similar situation exists today at Anshan, the iron and steel capital of China, but there the ore is brought to the complex by a 15XX volt dc electric circular railway from which branches radiate to the mines. Interestingly, most of Anshan's fleet of electric locomotives are much older than the thirty steam locomotives employed within the complex. Anshan provides a dramatic example of railway operation; at its height, the complex produced over 13 million tonnes (tons) of steel a year. Sixty different factories make up the site, including ten blast furnaces, three steel mills, a sinter plant, a huge coking plant, twenty rolling mills, two power stations, a refractory and machine repair shops. Ninety per cent of the ore is locally mined. The other ten per cent – a different grade for making specific types of steel – is brought in by main-line railway. The steam locomotives also move coal, scrap metals, limestone and magnesium in addition to the vast range of materials needed to keep the complex in good repair. Anshan continues to be active and this complex, along with others in China, will ensure that steam will have continued to work in partnership with iron and steel making for two centuries.

OPPOSITE LEFT: One of China's rare industrial Prairie 2-6-2s, in the form of YJ Class No. 291, basking amid the dense industrial haze of the Anshan complex.

OPPOSITE RIGHT: A Glasgow-built Sharp Stewart 0-6-0ST draws a rake of steel bars at Cosim works near Sao Paulo, Brazil.

ABOVE LEFT: The works plate of the Sharp Stewart 0–6–0ST at Cosim.

ABOVE RIGHT: Turkey's iron and steel works are located at Karabuk in the mountains near the Black Sea coast, and here one of the complex's Hawthorn Leslie 0-6-0STs empties molten slag from the blast furnaces down the slag bank.

LEFT: Anshan's steel mill number 1, with its bank of open-hearth furnaces in full cry, resembles a scene from the early 19th century.

Industrial Workhorses

Goods Carriers

The steam locomotive's pre-eminence as a mover of goods over the one-and-three-quarter centuries of its existence meant that it played a vital role in almost all industrial activities and, to this day, it remains in industrial locations throughout the world, albeit in ever-dwindling numbers. Industrial engines often had a longer life than their main-line relations, and some industrials are in excess of one hundred years old.

The development of coal-fired power stations meant that locomotives had to bring coal from the reception sidings with the main-line railway to the hoppers from which the furnaces for the power station boilers were fed.

Docks around the world also used fleets of locomotives to move imports and exports of rail-borne goods between the exchange sidings with the main-line railway and the loading quays.

As the Industrial Revolution spread, the vast quantities of bricks used in building led to a wide diversity of railways being developed, especially in the movement of clay from surrounding pits into the plants where the bricks were made. Although trucking, conveyor belts and overhead cables have rendered many rail systems obsolete, a classic example still exists at Ledo in upper Assam, an area in which 0.6m (2ft) gauge Bagnall Saddle Tanks have been active for over a hundred years.

Quarries for all kinds of raw materials have proliferated worldwide, particularly for iron ore and stone, but alternative handling techniques have decimated the traditional internal railways.

FAR LEFT: This standard 0-4-0ST with 35 cm (14 in) diameter cylinders was built by Andrew Barclay of Kilmarnock, and was one of two identical engines employed at Goldington power station south of Bedford. Both engines were converted to burn oil.

LEFT: The Calcutta Port Trust used a fleet of standard 1.7 m (5 ft 6 in) gauge 0-6-2 side tanks for operations around the docks. The class totalled 45 engines, the first of which came from Hunslet's works in Leeds in 1945. Subsequent batches were delivered from Henschel of Kassel and Mitsubishi of Japan.

ABOVE: A 0.6 m (2 ft) gauge Bagnall 0-4-0ST draws a rake of empty clay tubs out of Ledo brick works on the coalfield in upper Assam, bound for the clay pit.

RIGHT: One of the world's last steam-operated quarries was at Paso de los Toros in Uruguay, where this 0.6 m (2 ft) gauge Orenstein and Koppel 0-4-0WT worked on a half-mile stretch of line for more than fifty years bringing stone up to the main line for use as track ballast.

Heavy haulers

Sankong Bridge

Sankong Bridge in Harbin, the capital of Heilongjiang Province in north-eastern China, is the greatest train watching place in the world. A modest structure, it overlooks the marshalling yards of this important railway junction. A peep over the parapet rolls back the clock fifty years. Twelve steam locomotives may be seen scattered throughout the yard, their exhausts rising skywards amid a sea of wagons which stretch as far as the eye can see.

The bridge is located between two yards, one of which makes up the formations for the north-bound trains and the other for the south-bound trains. The bridge is actually built over the humping lines with wagons rolling from the reception sidings in either direction. Wagons of many different types, bearing all conceivable merchandise, roll down the humps, first through the King point, then the Queen points, before crossing primary and secondary retarders; dramas set against the amplified voice of the yard controller. Shunting activities are interspersed with accommodating transfer freights and through freights, all of which culminates in a steam movement beneath the arches at least once every two minutes. The *pièce de résistance* is the majestic departure of the south-bound freight train behind two QJs.

Above: A beautifully trimmed QJ class 2-10-2 heads a transfer freight.

Above right: A brace of China Railways' standard QJ Class 2-10-2s heads a south-bound freight away from the huge marshalling yard at Harbin.

Opposite : A China Railways' standard JS Class 2-8-2 Mikado starts a heavy south-bound freight out of Harbin amid winter temperatures of -30°C.

Above left: This is a typical view from Sankong Bridge – a blue trimmed QJ Class 2-10-2 backs down towards its train as a JF Class 2-8-2 Mikado performs a shunting movement.

Above right: Until the late 1980s, hump shunting at Harbin was done by JF 2-8-2s. These are pure American Light Mikes. Although the last examples were built in China as late as 1957, they are the descendants of engines delivered from the American Locomotive Company in 1918 for the South Manchurian Railway. Over 2,000 JFs were built, but today only a handful remain.

建设 5036

HEAVY HAULERS

ENGINES ON SHED

ONE OF THE STEAM LOCOMOTIVE'S PRINCIPAL DISADVANTAGES IS ITS NEED FOR FREQUENT SERVICING – COALING, WATERING, ASH RAKING, BOILER WASHING, LIGHTING UP AND TURNING – WHICH NECESSITATES FREQUENT VISITS TO THE SHEDS. THIS GREATLY REDUCES AN ENGINE'S AVAILABILITY AND ALSO MAKES STEAM TRACTION LABOUR INTENSIVE. ALTHOUGH THESE ACTIVITIES WERE UNDENIABLY FASCINATING AND THRILLING TO WATCH, THAT HELD LITTLE CREDENCE WITH FINANCIAL DEPARTMENTS WHICH DEMANDED MAXIMUM AVAILABILITY AT LOWEST COST IN ORDER TO COPE WITH COMPETING MODES OF TRANSPORTATION.

LEFT: A former Buenos Aires and Great Southern Railway 2-8-0, in the twilight of its life, at Olavarria depot. One hundred of these 1.7m (5 ft 6 in) gauge engines once worked across Argentina's fertile pampas.

TOP: An Indian Railways, Canadian-built, CWD Class 2-8-2 is coaled by a mobile steam crane at Bandel depot in Bengal. The engine is "blowing down" to eject scum and chemical impurities from its boiler.

ABOVE: China Railways' QJ Class 2-10-2s are serviced between turns of duty at Harbin motive power depot.

ABOVE: Two QJ Class 2-10-2s receive intensive servicing on a dull winter's evening at Harbin. These activities, which go on night and day, include cleaning, oiling, ashpan raking, coaling, watering, filling of sandboxes and turning. At its peak, Harbin had over 120 locomotives allocated.

LEFT: An Indian Railways metre (3 ¼ft) gauge YG Class 2-8-2 Mikado at Wankaner, the country's last main-line steam shed, early in 1999. The boiler washer, seen inside the smokebox, pauses between duties, having just shovelled a prodigious quantity of char from the smokebox as part of preparing the engine for its weekly boiler wash-out.

NEAR LEFT: Another Indian Railways YG 2-8-2 at Wankaner, with the boiler wash-out in progress. High-powered water jets are coursed through the boiler, and the tubes are manually scraped through inspection holes to prevent chemical deposits from furring them up. If left untreated, these deposits would greatly impair the engine's steaming capability.

Heavy Haulers

Indian Giants

India's magnificent XE Mikados were the most powerful conventional steam locomotives ever to work on the sub-continent. They were introduced on the 1.7 m (5 ft 6 in) broad-gauge lines, and were one of the famous X series of standards which were the next generation from the BESA standards of 1903. The XEs totalled 58 engines, built by William Beardmore of Dalmuir on the Clyde, and at the Vulcan Foundry, Newton-le-Willows, Lancashire, for heavy coal hauls on the East India Railway.

The XE Mikados spent half a century hauling 2,000 tonne (ton) coal trains over the hill regions of Bengal, particularly on the section between Dhanbad on the eastern coalfield and Jha Jha. However, this powerful British Mikado thrived only briefly in Britain, in absolute contrast to America, where the type was heavily favoured both at home and for export. America built more Mikados than any other form of locomotive.

Britain's indigenous Mikados were all produced by Gresley of the LNER and totalled eight engines – two P1s and six P2s. The XEs were very similar to Gresley's 1925 P1s and, despite the XE having two cylinders compared with the P1's three, the power of the two designs was roughly the same. More British Mikados were envisaged, not least in the preparation of British Railways' standard designs of 1951, but in the event, the 2-10-0 was chosen.

The classic British lineage of the XE rendered it of tremendous historical importance, not least as the two P1s were withdrawn in 1945, by which time the P2s had been converted into Pacifics. It is amazing to think that for over fifty years after the P1's demise, almost identical engines remained active in India.

During the late 1970s, most XEs were condemned by Indian Railways, but a few were given major overhauls at the Jamalpur works and sold to private industry, notably cement and thermal power stations. Their survival in these environments has been largely undocumented for over twenty years.

ABOVE: XE Class No. 22523, built by the legendary Clydeside shipbuilder William Beardmore of Dalmuir in 1930, noisily propels a rake of coal empties along the yard at Ghoradongri thermal power station in Madhya Pradesh in January 1989.

LEFT: XE No. 22523 heads a rake of empty wagons from Ghoradongri thermal power station to Indian Railways' exchange sidings.

OPPOSITE LEFT: The last working XE was No. 22502, one of the Beardmore engines of 1930. It survived, working between Korba thermal power station and Manikpur colliery in Madhya Pradesh, until 1997, by which time it had become the last large conventional British steam locomotive left in world service. This photograph was taken in January 1997.

OPPOSITE CENTRE: With its Indian Railways number plate missing, a hastily improvised alternative was adopted.

OPPOSITE RIGHT: No. 22523's tender sprang a leak which appears never to have received any attention

HEAVY HAULERS

CLASSIC FORMS

THE EVOLUTION OF THE FREIGHT LOCOMOTIVE WAS DICTATED BY TWO PRIMARY FACTORS: FIRSTLY, THE NEED FOR INCREASINGLY POWERFUL LOCOMOTIVES TO WORK HEAVIER TRAINS AS INDUSTRIES DEVELOPED AND, SECONDLY, AS THE 20TH CENTURY DAWNED, TO SPEED UP HEAVY, SLOW-MOVING TRAINS WHICH WERE CLOGGING THE RAILWAY NETWORKS. A FURTHER DEMAND FOR FASTER FREIGHT TRAINS OCCURRED ONCE ROAD TRANSPORTATION BECAME A THREAT FROM THE 1920S ONWARDS.

One of the earliest and best-loved forms of freight locomotive was the 0-6-0, which appeared in Britain during the 1830s, the last examples being built as late as 1942. The progression from 0-6-0 to 0-8-0 and 0-10-0 was rapid in Europe and Russia, where the 2-8-0s also appeared; 2-10-0s were in operation before 1920. The most common freight locomotive ever to run was the Russian E Class 0-10-0, of which some 13,000 examples were built.

In America and Africa, leading and trailing axles were more common, especially when track conditions were less than perfect. North America favoured 2-6-0, 2-8-0, 2-8-2, 2-10-0 and 2-10-2s while in Africa and Latin America, 4-8-0, 2-8-2 and 4-8-2s were widely used. The ultimate freight haulers in North America were the 4-8-2, 4-8-4 and 2-10-4s, while the big compound Mallets and articulateds, which culminated in the 520 tonne (512 ton) Big Boys, had no equal.

OPPOSITE LEFT: *One of the world's first 0-8-0 heavy goods engines lies abandoned at Olloneigro Colliery in Spain. It was built by Hartmann of Chemnitz in 1879 and was formally named "El Cavado".*

OPPOSITE RIGHT: *Classic freight engines in the form of Austrian Empire designed 0-10-0s and 2-10-0s, contrasting with the classic American S160s of World War II, lie abandoned in the huge locomotive dump at Thessaloniki in northern Greece.*

RIGHT: *A former Indian Railways AWE Class 2-8-2 heads a rake of empty wagons back to the colliery at Manikpur from Korba power station.*

FAR RIGHT: *This is the view from the footplate of a Finnish Railways TK3 2-8-0 engaged on tripping duties around Rovaniemi on the Arctic Circle.*

BELOW: *A Russian LV Class 2-10-2 clearly reveals its American ancestry. Introduced by the Voroshilovgrad works in 1952, building of these engines was cut short by the decision to abandon steam.*

Fleet of Foot Flyers

These days, steam trains may be regarded as an outdated, plodding form of transport, but a century ago, three-figure speeds were being reached. Very few diesels the world over have bettered steam's top speeds. Almost a hundred years ago, steam trains were running between London and Birmingham in two hours; this could not be equalled by motorway or highway today, while in 1999 the fastest Virgin electric train took one hour thirty minutes – a gain of just thirty minutes in almost a hundred years.

Britain's Great Western Railway featured dramatically in early high speed achievements. In 1903 their "City of Truro" reached a speed of 164 kph (102 mph) during the descent of Wellington Bank, breaking the three-figure speed barrier. In 1906 the company's new "Saint" Class No.2903, "Lady of Lyons", reached 193 kph (120 mph). Then, in June 1932, the GW's "Cheltenham Flyer" ran the 124 km (77.3 miles) from Swindon to London in 56 minutes 47 seconds – a start to stop average of 131 kph (81.68 mph).

Left: The China Railways SL7 Class Pacific was built in the streamlined tradition of the 1930s for working the Asia Express between Dalian and Mukden.

Right: A North Eastern Railway Z1 Atlantic caught on film with a monoplane during the early years of the 20th century.

Above: These magnificent Atlantics hauled the Milwaukee Road's Hiawatha over the 663 km (412 miles) between Chicago and St Paul at an average speed in excess of 129 kph (80 mph).

16

Fleet of Foot

Pacifics & Mikados

The Pacific will go down in history as the Golden Mean of express engines. Its reign on the world's railways spanned one hundred years, and some of the most striking and successful steam locomotive designs have been Pacifics. It had a beautiful wheel arrangement when combined with large diameter driving wheels. Innumerable locomotive designers have, with the boldness and verve of great composers, produced the most delightful variations on the Pacific theme.

Streamlining added to the Pacific's allure, and doubtless made a contribution to the British A4's record-breaking run on 3 July 1938, when locomotive No. 4468 "Mallard" reached 203 kph (12 mph) between Grantham and Peterborough, establishing a world speed record for steam traction which is still unbroken. Traditionally, express passenger engines had a shorter life than the lower minions in the locomotive order; they were both highly specialized and covered longer daily distances. Also, the more prestigious passenger services were early candidates for diesel power once modernization programmes began, and as a result, the Pacific declined rapidly following World War II. By the 1970s it had disappeared completely from many areas. In contrast, the Mikado was primarily a freight locomotive, although some potent passenger examples did appear in North America. During the final years of steam development, the Mikado evolved into a mixed-traffic engine, as epitomized by the French 141Rs

ABOVE: A young Bengali herdsman sits on the banks of a Ganges tributary as one of Indian Railways' last XC express passenger engines ekes out its days on the Bolpur pick-up goods. The thoroughbred was working from Burdwan depot where the last surviving members of this class were allocated.

OPPOSITE LEFT: One of Indian Railways' last metre (3¼ ft) gauge YGs heads a Morbi to Wankaner train through the grassy, light-dappled cuttings near Vaghasia.

OPPOSITE CENTRE: Indian Railways' 1.7 m (5 ft 6 in) gauge WP Pacific No. 7247 prepares to leave its home depot at Asansol. The engine is specially decked for Indian Railways' annual locomotive beauty contest.

OPPOSITE RIGHT: The PT47 passenger-hauling Mikados were extremely handsome and of classic 1930s appearance; they epitomized the Polish school of locomotive design.

Fleet of Foot

A Century of Evolution

The evolution of the steam locomotive was dictated by the constant demand for more power and speed. In Britain and areas of Europe, the 2-4-0, which emerged as the principal type of locomotive in the 1840s, quickly grew into the 4-4-0 and subsequently the 4-6-0. The need for increasingly larger fireboxes produced, in turn, the Pacific and the Mikado. Traditionally, Pacifics hauled the fastest expresses, while the passenger-designed Mikados were ideal for heavy trains on less exacting schedules over routes which were graded and had axle weight limitations.

OPPOSITE LEFT: *This delightful 4-4-0 was built at Sharp Stewart's works in Glasgow in 1892 for working passenger trains over Brazil's metre (3¼ ft) gauge Mogiana Railway.*

OPPOSITE RIGHT: *The world's last inside cylinder 4-4-0s – the classic express engines of late Victorian Britain – ended their days in the Pakistan Punjab with the beautiful SPS Class which had 1.9 m (6 ft 2 in) diameter driving wheels.*

ABOVE: *The last surviving Indian Railways XC express passenger engine was built by the Vulcan Foundry in 1928.*

RIGHT: *The last surviving Indian Railways XB express passenger engine ended its days working from Rajahmundry depot in Andhra Pradesh.*

FAR RIGHT: *The Indonesian State Railways' B50 Class 2-4-0s were the last examples of this early passenger wheel arrangement to survive. The first B50s were built by Sharp Stewart in Manchester in 1880.*

SUBURBAN & BRANCH BEFORE ELECTRIFICATION

AS A RESULT OF THE INDUSTRIAL REVOLUTION, CITIES AROUND THE WORLD EXPANDED RAPIDLY, MAKING FASTER URBAN TRANSPORT ESSENTIAL, AND MANY DYNAMIC SYSTEMS EMERGED; LONDON'S UNDERGROUND AND THE OVERHEAD RAILWAYS IN NEW YORK AND CHICAGO ARE OBVIOUS EXAMPLES. MANY OF THE MAJOR SYSTEMS BECAME EARLY CANDIDATES FOR ELECTRIFICATION BECAUSE OF THE POLLUTION CAUSED BY STEAM LOCOMOTIVES PASSING THROUGH TUNNELS AND AMID DENSE URBAN CONURBATIONS. BY CONTRAST, BRANCH LINES WERE OFTEN RURAL AFFAIRS AND LEISURELY IN THEIR MODE OF OPERATION. MANY LINES REMAINED STEAM OPERATED THROUGHOUT THEIR EXISTENCE, SOME WITH FORMER SUBURBAN ENGINES MADE REDUNDANT BY ELECTRIFICATION.

Suburban steam trains are now extinct and in many places where they flourished their lifespan was limited. Most of the world's first railway electrification schemes were concerned with suburban networks. There were two principal reasons; firstly, the elimination of pollution and secondly, it was widely perceived that electric traction would provide quicker starts and faster speeds to enable a more rapid flow of trains on sections which, in many cases, were very densely operated. In America, the famous Chicago and New York elevateds, with their huge numbers of Forney tank engines, were electrified in the early years of the 20th century. In Britain, electrification of London's Metropolitan and District underground lines was completed by 1905, and as early as the 1930s, extensive parts of Britain's Southern Railway, serving the dense commuter belt south of London, were converted to 750 volt dc, third-rail operation. Similar patterns occurred in most parts of Europe although steam suburbans did survive in Paris until the late 1960s and the Victorian-style metre (3¼ ft) gauge network around Oporto in Portugal continued in full-blooded operation into the 1970s. Most steam suburban engines were tanks, which were ideal for running in either direction. Many of the designs were extremely fleet-footed and often powerful enough to enable heavy commuter trains to operate at suitable line speeds and so minimize potential delays to faster, long-distance trains.

As suburban railways have prospered around congested cities of the world, the fate of branch lines has been less happy. These were seldom electrified and the rise of road transportation in the form of motor cars, buses and trucks rendered many secondary lines allegedly uneconomic. Closures have been legion; not just in the developed nations but all over the world, including some of the poorest countries.

TOP LEFT: Builder's plate from an Indian Railways metre (3¼ ft) gauge 4-6-0.

TOP: Kent miners await their evening train home, headed by a former LMS Fairburn 2-6-4T.

ABOVE LEFT: A Portuguese metre (3¼ ft) gauge 0-4-4-0T Mallet at Oporto Trindade station.

ABOVE: A Yugoslav State Railways 51 Class 2-6-2T at Vrginmost on the branch line between Karlovac and Sisak.

OPPOSITE: One of thirty WT Class 2-8-4 suburban tanks built by India's Chittaranjan locomotive works between 1959 and 1967. Here, No. 14011 is seen at Rajamundry from where it worked cross-country passenger trains.

14011
WT

BRANCH LINE ENGINES

LOCOMOTIVES WHICH WERE SUITABLE FOR SUBURBAN OPERATION WERE OFTEN IDEAL FOR MOVING ON TO BRANCH LINES ONCE THE URBAN NETWORKS HAD BEEN ELECTRIFIED. OTHER DESIGNS WERE BUILT FOR BOTH PURPOSES. BRANCH LINES WERE ALSO TRADITIONAL HAUNTS OF DOWN-GRADED MAIN-LINE ENGINES, WHICH HAD BECOME EITHER TOO OLD OR TOO WEAK, OR BOTH, FOR THEIR ORIGINAL PURPOSE.

Some interesting forms of specialized motive power have occurred on branch lines in the form of push and pull auto operation, in which a set formation of locomotive and coaches could be driven from either end. The introduction of diesel multiple units on to secondary routes had its origins in the steam railcars, which operated with varying degrees of success from the 1920s onwards. The introduction of diesel railcars on to branch lines was intended to attract more travellers, thus producing economies in operation and preventing closure of marginally remunerative routes. In reality, many closed as road transportation was allowed to develop in an unfettered way.

Few branch lines remain steam worked today; exceptions are some metre (3¼ ft) gauge lines in the former German Democratic Republic and in Myanmar, and the celebrated 0.8 m (2 ft 6 in) gauge Esquel branch line in Patagonia.

LEFT: A Burma Railways metre (3¼ ft) gauge 2-6-4T at the ash pits. Local people scour the track for unburned pieces of fuel.

ABOVE: The LB&SCR Terrier 0-6-0Ts had characteristic chimney and safety valves. These engines were introduced in 1872 to cope with South London's rapidly growing suburban traffic.

OPPOSITE ABOVE: Bullocks bring cotton to Pulgeon Mill as the morning passenger train departs to Arvi behind an Indian Railways 0.8 m (2 ft 6 in) gauge ZP Class Pacific.

OPPOSITE BELOW LEFT: In a 1948 scene at Chesham, a former Great Central C13 4-4-2T operates the shuttle up the main line to Chalfont and Latimer.

OPPOSITE BELOW RIGHT: This CNL7 4-6-0, No.767, pictured at Lockport Penna on 13 October 1941, is equipped with a Wooten firebox for consuming low-grade anthracite coal.

67438

767

SUBURBAN & BRANCH

INDIAN BYWAYS

THE INDIAN SUB-CONTINENT WAS BLESSED WITH AN INCOMPARABLE NETWORK OF BRANCH AND COUNTRY RAILWAYS. MANY OF THE SMALLER RURAL LINES WERE 0.8 M (2 FT 6 IN) GAUGE AND WERE WORKED BY AN INCREDIBLE ARRAY OF VINTAGE LOCOMOTIVES, ALMOST ALL OF WHICH WERE BRITISH-BUILT. THESE RAILWAYS WERE THE LIFELINES OF MANY RURAL COMMUNITIES AND CONNECTED TO LONGER DISTANCE METRE (3¼ FT) OR BROAD-GAUGE NETWORKS.

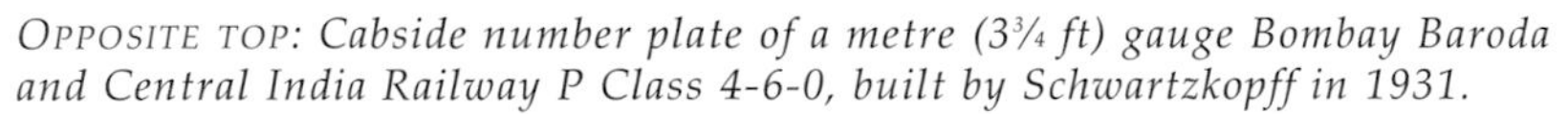

Opposite top: *Cabside number plate of a metre (3¼ ft) gauge Bombay Baroda and Central India Railway P Class 4-6-0, built by Schwartzkopff in 1931.*

Opposite: *This is Pargothan on Indian Railways' 0.8 m (2 ft 6 in) gauge country line from Pulgaon to Arvi. The engine is a ZP Class Pacific.*

Above: *The Darjeeling Himalayan Railway's famous B Class 0-4-0STs were introduced by Sharp Stewart of Glasgow in 1899.*

Above right: *Burdwan to Katwa Railway's No. 3, a 0.8 m (2 ft 6 in) gauge 0-6-4T built by Bagnalls of Stafford in 1914, arrives at Burdwan.*

Right: *Passengers waiting at Sorta station for train "643 down", which left Pulgaon for Arvi daily at 08.00.*

SHUNTERS & BANKERS

MARSHALLING ENGINES

THE SHUNTING LOCOMOTIVE EVOLVED ONCE RAILWAY SYSTEMS BECAME INTERCONNECTED AND MARSHALLING YARDS WERE CREATED FOR ASSEMBLING TRAINS BOUND FOR DIFFERENT DESTINATIONS. UNTIL THIS TIME, MAIN-LINE ENGINES UNDERTOOK WHATEVER STATION OR YARD MOVEMENTS WERE NECESSARY. BY THE MIDDLE OF THE 19TH CENTURY MANY OBSOLETE MAIN-LINE TYPES WERE AVAILABLE TO BE DOWNGRADED AS SHUNTERS. HOWEVER, THEY WERE INVARIABLY ILL-SUITED TO THE TASK AND ONCE MARSHALLING YARDS DEVELOPED, SUITABLY DESIGNED LOCOMOTIVES BECAME NECESSARY.

TOP LEFT: One of Argentina's 1.7 m (5 ft 6 in) gauge General Roca Railway BE Class 0-6-0STs shunts at Bahia Blanca. These typical British shunters were introduced by the North British of Glasgow in 1904 with some later examples appearing from Kerr Stuart of Stoke-on-Trent.

TOP RIGHT: A former United States Army Transportation Corps (USATC) 0-6-0T, built in America for operations during World War II, takes on water at Drama in northern Greece.

ABOVE RIGHT: The number plate of one of the USATC World War II 0-6-0Ts, classified XK2 by China Railways.

RIGHT: An early French Nord Railway 0-6-0 shunter.

OPPOSITE: These two former Great Indian Peninsular Railway bankers were built for pushing trains up the Ghats out of Bombay. They are seen here at the Hindalco Aluminium smelter in Renukut. On the right is an 0-8-4T North British of Glasgow 1920 and, left, a 2-8-4T North British of Glasgow, 1907.

Shunting locomotives began to appear on main lines in the second half of the 19th century, having long been established in industrial environments. The compactness of tank engines made them ideal for the job; the weight of water bearing directly on the driving wheels gave essential adhesion and the absence of a tender facilitated easy running in either direction. In many cases, tenders were not essential as coal and water could be taken on frequently between bouts of shunting. In general, shunters had no need for leading or trailing axles; operating speeds were low and axle loading restrictions hardly existed in most sidings and yards. By far the most common shunting tank was the 0-6-0 with tanks of side, pannier, Saddle or Well variety. The 0-4-0 was also widely used, especially for work in restricted areas or for negotiating very sharp curves.

In North America, 0-4-0 and 0-6-0s appeared in both tender and tank varieties. By the 20th century, shunters around the world began to increase in size, not least in North America where some huge ten-wheeled examples were used.

L-2
L-3

SHUNTERS & BANKERS OF ALL GAUGES

AS THE TWENTIETH CENTURY PROGRESSED, THE STEAM SHUNTING ENGINE PROPER BEGAN A RAPID DECLINE FOR TWO DIFFERENT REASONS. FIRSTLY, HUGE NUMBERS OF OLD GOODS ENGINES WERE AVAILABLE FOR DOWN-GRADING AND, SECONDLY, THE FIRST INROADS DIESEL MADE INTO STEAM SUPREMACY WAS IN THE SHUNTING YARDS. A DIESEL HAS EXCELLENT TORQUE FOR MOVING HEAVY LOADS FROM A STAND WHEREAS STEAM SHUNTERS ALMOST ALWAYS SLIP VIGOROUSLY. FURTHER ECONOMY OF OPERATION WAS MADE BY THE ABILITY OF A DIESEL TO SHUT DOWN BETWEEN BOUTS OF WORK.

Hump shunters propelled long rakes of wagons up a steep slope to the summit from which they would roll by gravity down the other side and on into the sidings. Pushing heavy rakes of wagons up a steep gradient was no task for four- or six-wheeled engines and humpers in Britain were often 0-8-2 or 0-8-4, with ten- or twelve-wheeled examples appearing in North America. Banking engines were close relations of humpers; many bankers would spend their lives on one stretch of railway, pushing trains to the summit before rolling down light engine and waiting for the next train to come up.

Among the most famous bankers was the Midland Railway's "Big Bertha" of 1919 – Britain's only 0-10-0, built for pushing trains up the 1 in 37 Lickey Incline south of Birmingham. Even more stupendous was the Baltimore and Ohio's "Old Maud", an 0-6-6-0 four cylinder compound Mallet of 1904, which was North America's first giant articulated and acted as a catalyst for the world's biggest and most dramatic locomotives.

ABOVE: *Asmara loco shed in Eritrea with Breda 0-4-0 Well Tanks.*

RIGHT: *Revolutionary zeal on a Cuban Alco. "Venceremos" means "We shall overcome".*

LEFT: *An 0-4-0 Camel Back shunter built by Baldwin's in 1902 for the Reading Railroad, is seen at Rutherford on 7 May 1939.*

OPPOSITE RIGHT: *Former Lancashire and Yorkshire Railway 0.5 m (1 ft 6 in) gauge works shunter 0-4-0ST "Wren" was one of eight engines built between 1887 and 1901.*

OPPOSITE LEFT: *This Baldwin-built switcher 0-6-2ST was built in 1896 for Brazil's 1.6 m (5 ft 3 in) gauge Paulista Railway.*

UNCONVENTIONAL DESIGNS

ARTICULATEDS

THE FAIRLIE, KITSON MEYER AND GARRATT WERE ARTICULATED VARIANTS FROM THE CONVENTIONAL STEAM LOCOMOTIVE AND WERE NECESSARY WHERE RELATIVELY POWERFUL ENGINES WERE NEEDED TO HAUL HEAVY TRAINS OVER TRACKS WHICH WERE TIGHTLY CURVED, STEEPLY GRADED OR LIGHTLY LAID. THEY OPERATED IN MANY PARTS OF THE WORLD, ESPECIALLY THE GARRATT, WHICH WAS MOST ASSOCIATED WITH THE ROUGH TERRAIN OF AFRICA, WHERE THE WORLD'S LAST EXAMPLES CAN STILL BE FOUND IN SERVICE.

Robert Fairlie patented a double-bogied, double boiler locomotive which became an important articulated type from its conception in 1863 up to World War I. The majority of the engines took the form of 0-4-4-0Ts or 0-6-6-0Ts. One advantage was a firebox unrestricted by frames and wheels, but the amount of coal and water which could be carried was limited, especially as the designs, inevitably, grew larger.

In contrast, the Kitson Meyer, which also had its firebox between the wheels, consisted of a large tank engine mounted on double-power bogies, with cylinders often placed at the outer ends of each bogie. This invariably resulted in the steam from the rear cylinders passing through the rear water tank and being exhausted through a separate chimney. Though the Kitson Meyer was a popular articulated, its advance was cut short by the more successful Garratt.

On a Garratt engine, the boiler and firebox were free of axles and so could be built to whatever size was needed, both for ample generation of steam and for combustion of gases, by the provision of a deep firebox. Conversely, as the wheels were free of the boiler, they could be made to whatever diameter was considered best. By placing the engine's wheels and cylinders under a front water unit and rear coal unit – situated on either side of the boiler – the engine's weight was spread over a wide area. Simply by articulating these two units from the boiler, a large powerful locomotive, capable of moving heavy loads over curved, graded or lightly laid lines, could be built. The Garratt was the most successful type of articulated steam locomotive and some 2,000 were built for gauges ranging from 0.6 m (2 ft) to 1.7 m (5 ft 6 in).

ABOVE LEFT: A brace of South African Railways' mighty GMA 4-8-2+2-8-4 Garratts departs from City View with freight on 18 June 1974. The water tank between the locomotives spreads the axle weight enabling these powerful engines to operate over lightly laid track.

ABOVE RIGHT: The name plate of East African Railways' 4-8-2+2-8-4 Mountain Class Garratt No. 5928. These engines worked the 534 km (332 mile) line between Mombasa on the Indian Ocean and the Kenyan capital Nairobi.

BELOW LEFT: A works profile of one of the 4-8-4+4-8-4 Garratts, built by Beyer Peacock for the New South Wales Railway, Australia.

BELOW RIGHT: The Fairlie was an early form of articulated. Some of the first double Fairlie locomotives worked on the 0.6 m (1ft 11½ in) gauge Ffestiniog Railway in North Wales, the first ones appearing as early as 1869.

OPPOSITE: The last surviving 0-6-6-0 Kitson Meyer ending its days at Taltal on Chile's Pacific coast.

Unconventional Designs

Rare Survivors

The steam locomotive's history is studded with amazing variations on the common theme. Its premier role in world transportation throughout the 19th and much of the 20th centuries created a wide diversity of demands to which creative engineers could apply their talents. Some variations were farcical and never got off the ground, but Sentinels and Shays proved their worth, as did the Fireless, which was akin to a thermos flask on wheels. In contrast, the monorail had little practical success, although its designer's intention was to revolutionize rural transport.

One of the most interesting variations on the conventional locomotive was the Sentinel, which had a vertical boiler, with a high working pressure of around 275 pounds per square inch; the cylinders were positioned vertically and transmission was geared. Sentinel claimed huge savings in running costs compared with the "inefficiencies" of the conventional industrial engine by virtue of the superior boiler and transmission. Following its introduction in 1923, some 850 were built, the majority of which were for export to all parts of the world.

The Sentinel's *raison d'être* was efficiency but the Shay – the brainchild of a backwoods logging engineer – was specifically designed to haul heavy lumber trains over temporary tracks which were frequently engulfed in mud and abounded in curves and gradients. Shays had their wheels mounted in articulated trucks which provided flexibility and spread the engine's weight. The vertical cylinders, which were placed on the engine's right-hand side, drove a horizontal shaft which connected to all wheels by means of pinions slotted into bevel gears. The crank shaft was made flexible by the insertion of universal joints. The geared transmission provided an easy turning movement and enabled operation in the most appalling track conditions. The Shay was the classic lumber engine of the American Pacific north-west and was adopted by Lima of Ohio.

The Fireless evolved to fulfil two distinct requirements in operation; it was ideal for those industries firstly with a ready supply of high-pressure steam, and secondly where sparks from a conventional engine could wreak havoc, such as paper mills, jute mills or explosives factories. One charge of high-pressure steam from the work's boilers enabled the Fireless to shunt around the plant for several hours before needing a recharge; there was no need for either a fireman or fuel and there was little to go wrong mechanically. Arguably, the Fireless was the most efficient shunting unit ever devised for industries with steam on tap unless, of course, it ran out of steam at the opposite end of the works from the factory boilers. However, because of the all-pervading witchhunt against steam, even the trusty Fireless has disappeared from most parts of the world.

Opposite above: One of the world's last surviving Sentinels is to be found at a Brazilian wagon works in Sao Paulo State.

Opposite below: An old Lima two-truck Shay raises steam at dawn in readiness for a day's work around the saw mill of the Insular Lumber Company on the Philippine island of Negros.

Above left: Ireland's Listowel and Ballybunnion Railway operated Lartigue's monorail system which used double-engined vehicles straddling a trestle rail. The system operated between 1888 and 1924.

Above right: A Baldwin-built 0-4-0 Fireless of 1917 operates at Bolivia Sugar Mill in Cuba.

Left: This delightful Crane Tank was worked at an Indian sleeper depot and was built by Manning Wardle of Leeds in 1903. The crane structure and auxiliary engine came from the nearby works of Joseph Booth and Son in the Leeds suburb of Rodley.

Unconventional Designs

Steam Trams

THE SNOWDON MOUNTAIN RAILWAY, WITH ITS RACK AND PINION, MADE A MARKED CONTRAST TO THE STEAM BRAKE ENGINES WHICH PLIED THE SANTOS JUNDIAI IN BRAZIL. THE STEAM TRAM PROPER, THOUGH SIMILAR IN APPEARANCE, HAD A TOTALLY DIFFERENT FUNCTION FROM EITHER OF THESE. IT WAS USED TO WORK AROUND STREETS AND DOCKS. THE STEAM RAILCAR – THE FORERUNNER OF TODAY'S UBIQUITOUS DIESEL MULTIPLE UNITS – WAS EMPLOYED ON BRANCH LINES AND CROSS-COUNTRY ROUTES IN THE 1920S AND 1930S.

Some mountain and hill railways continue to be steam worked, and in recent years new locomotives have been built for them. Some systems, like the Snowdon, have used steam since the lines were opened in the last century. It is sad that the unique steam brake vans which plied the cable portion of the Santos Jundiai in Brazil are now just a memory. Throughout the climb up the escarpment, these engines moved wagons along the intermediate level sections set between the gradients. The system was cable worked and trains going up were balanced by trains coming down.

Tram engines – in which all the moving parts were encased – appeared on roadside tramways and secondary routes which involved on-street running for part of the journey. They were especially prevalent in Holland and Java, once part of the Dutch East Indies.

Steam railcars were also encased and actually carried passengers. They operated in a variety of guises in many parts of the world. In Britain, they were often intended to compete with electric trams for suburban work, while on uneconomic country branch lines they proved simple to operate and often worked on the one engine in steam principal. The steam railcar is a perfect example of the relationship between suburban and branch-line engines.

OPPOSITE ABOVE: The historic 0.8 m (2 ft 7½ in) gauge Snowdon Mountain Railway in Wales opened in 1896. Here, locomotive No. 8 "Eryri", a 0-4-2T built by Swiss Locomotive Works of Winterthur in 1923, begins the steep climb to Clogwyn station.

OPPOSITE LEFT: The last steam tram in the world to remain in service lingers at a Paraguayan sugar mill. Built by Borsig of Berlin in 1910, the veteran originally worked through the streets of Buenos Aries, the capital of neighbouring Argentina.

OPPOSITE RIGHT: Number plate of one of the Kerr Stuart brake engines of 1900.

ABOVE: A unique cable railway of 1.6 m (5 ft 3 in) gauge operates up the escarpment between Piassaguera and Pindamonhagaba on Brazilian State Railways. The tram-like engines were built between 1900 and 1931 by Kerr Stuart and Robert Stephenson.

LEFT: One of the last surviving steam railcars lies out of use at Damazeen in the Sudan. It was built by Clayton of Lincoln.

EARLY FORMS

THE 0-4-0 AND 0-6-0 WERE EARLY FORMS OF LOCOMOTIVE; THE 0-4-0 WAS INTRODUCED IN THE 1820S AND THE 0-6-0 DURING THE 1830S. THEY WERE PRIMARILY BRITISH AND EUROPEAN CONCEPTS. IN CONTRAST, AMERICA'S EARLY MIXED-TRAFFIC ENGINES WERE PRIMARILY 4-4-0S; THEIR LEADING WHEELS PROVIDED STABILITY ON THE GENERALLY INFERIOR TRACK OF THE PIONEERING YEARS.

The 0-4-0 was an early form which thrived briefly and was rapidly overtaken by the 0-6-0, which provided better adhesion and became a principal mixed-traffic type built in vast numbers for more than a century. However, by an amazing quirk of circumstance, the 0-4-0 outlived its larger relation – two metre (3¼ ft) gauge examples named "Tweed" and "Mersey" were built by Sharp Stewart of Manchester in 1873 for India's Tirhut Railway. These and other members of their class worked mixed trains between Tirhut and Patna before being retired into sugar mill service where they are still active in their 125th year.

The inside cylinder 0-6-0 was a British speciality; many thousands were built to innumerable different designs, many virtually identical in overall specification. A lot of 0-6-0 designs were specifically for hauling slow, plodding freight trains but others – usually with slightly larger driving wheels – were excellent mixed-traffic engines, their modest size enabling them to traverse a wide variety of lines.

The world's first standard steam design was an inside cylinder 0-6-0 in the form of Ramsbottom's famous DX Class mixed-traffic engines of which 943 were built by the London and North Western Railway. The full versatility of the inside cylinder 0-6-0 was revealed at the end of their lives when many made a successful transition into the shunting yards.

ABOVE: This scene at Lucknow locomotive depot on India's Northern Railway, shows an inside cylinder 0-6-0 built in 1923 at Armstrong Whitworth's works on the Scotswood Road, Newcastle upon Tyne. Classified SGS, the engine is one of the famous BESA standards introduced for India's 1.7 m (5 ft 6 in) gauge lines in 1903. The inside cylinder 0-6-0 was one of the classic forms of British steam locomotive and the last survivors of the type ended their days on the Indian sub-continent.

RIGHT: This veteran 0-4-0 named "Mersey" works at Hathua Sugar Mill in northern India, and is one of the world's oldest steam survivors, having been built by Sharp Stewart in 1873.

OPPOSITE: Malakwal Junction in the Pakistan Punjab had a number of inside cylinder 0-6-0s until the late 1990s. They were employed on both freight and passenger workings on the network of secondary lines radiating from this country junction.

2505

JACK OF ALL TRADES

MIXED-TRAFFIC ENGINES

MOUNTAINS (4-8-2S), PRAIRIES (2-6-2S) AND MIKADOS (2-8-2S) WERE ALL PRE-EMINENT IN MIXED-TRAFFIC DESIGNS. THE 4-8-2 WAS PARTICULARLY WIDESPREAD IN SOUTH AFRICA, WHEREAS PRAIRIES WERE ABUNDANT IN AMERICA AND RUSSIA WHERE THE CELEBRATED SU CLASS TOTALLED SOME 3,750 EXAMPLES. THE UBIQUITOUS MIKADO BECAME A MIXED-TRAFFIC TYPE AROUND THE WORLD DURING THE LATER YEARS OF STEAM AS EPITOMIZED BY THE FRENCH 141RS WITH 1,340 ENGINES, INDIA'S WG CLASS WITH 2,450 EXAMPLES AND CHINA RAILWAYS' JS CLASS, WHICH TOTALLED UPWARDS OF 2,000 LOCOMOTIVES.

LEFT: A China Railways JS Class 2-8-2 heads a Yichun-bound passenger train away from Nancha in a winter temperature of -30°C. The banking engine is a QJ Class 2-10-2.

ABOVE: Eritrea's rebuilt railway system has brought back to life the network's Italian built 0-4-4-0 four-cylinder compound Mallets. Here No. 442.59 (1938) heads an engineer's train at Digdigta.

OPPOSITE: A South African Railways Class 14CR 4-8-2, built by ALCO in Montreal in 1919, heads the 13.15 Worcester to Riversdale train between Ashton and Bonnievale on 13 March 1976.

JACK OF ALL TRADES ON FOUR CONTINENTS

THESE PICTURES INDICATE THE GREAT DIVERSITY OF MIXED-TRAFFIC ENGINES, WHICH EMBRACED A WIDE RANGE OF WHEEL ARRANGEMENTS. MANY RAILWAYS THROUGHOUT THE WORLD OPERATED MIXED TRAINS, ESPECIALLY ON SECONDARY AND BRANCH LINES, WHERE SCHEDULES WERE SLOW AND THE COLLECTION AND SETTING DOWN OF FREIGHT WAGONS, PLUS WHATEVER SHUNTING WAS NECESSARY, WAS PERFORMED WHILE THE PASSENGERS WAITED.

The trend towards mixed-traffic designs meant easier locomotive diagramming, or administration, for the running departments, best availability and also greater standardization of types. Most passenger trains around the world were timed at average speeds well below 1.5 km (1 mile) a minute while, in general, freight speeded up as the 20th century progressed. Improvements in metallurgy and lubrication, along with better balanced design and improvements in track, all aided the engineers' ideal of locomotives suitable for a wide range of traffic.

It is an interesting paradox that at the end of steam development in Britain the twelve standard designs produced from 1951 for the newly formed British Railway were all mixed traffic, including the two Pacific designs, with the exception of the heavy mineral 9F 2-10-0. The 9Fs turned out to be one of the most successful mixed-traffic designs of all time, equally capable of hauling heavy mineral drags and working main-line expresses at speeds of up to 129 kph (80 mph)! On some sections, the engines enjoyed a brief spell utilizing top link power until it was forbidden by the operating authorities because of the enormous speeds of the moving parts, the driving wheel diameter of the 9F being only 1.6 m (5 ft 2 in).

Had steam continued, it is likely that the mixed-traffic engine would have become the operational norm, as was the case with the incoming diesels which were also able to run in tandem with only one crew, so enabling a whole new range of operational opportunities.

LEFT: This Belgian 2-6-0T, used for mixed-traffic work on secondary routes, was built by John Cockerill at Seraing, Belgium.

ABOVE: A perfect example of mixed-traffic operation on the Jersey City to Buffalo line shows identical members of the Delaware, Lackawana & Western Class 4-8-2, one pulling an express passenger train and one a mixed freight train. The freight train is being held in a loop to allow the passenger train to proceed in this classic North American scene from 1938.

OPPOSITE ABOVE LEFT: A typical mixed-traffic Mikado – a Turkish State Railways' Middle East 2-8-2. Two hundred of these engines were built for service during World War II for Britain's Ministry of Supply by Baldwin's, ALCO and Lima. They were shipped direct to Suez to work in Syria, Palestine and Persia.

OPPOSITE ABOVE RIGHT: East African Railways' Class 26 2-8-2 No. 2611, a 1952 Vulcan Foundry engine, heads a mixed train along the isolated line from Tabora to Mpanda in Tanzania in 1973. The engine is seen taking on water at Ugalla river.

OPPOSITE BELOW: One of Spain's RENFE Confederation Class 242F 4-8-4s takes on water at Castejon. Ten of these locomotives were built by Maquinista of Barcelona in 1955–6.

JACK OF ALL TRADES

FAMOUS DESIGNS

THE 4-6-0 AS A MIXED-TRAFFIC ENGINE WAS PIONEERED BY THE PRUSSIAN STATE RAILWAYS' P8 GENERAL UTILITY 4-6-0S OF 1906. AS A RESULT OF TWO WORLD WARS, THE P8S WERE WIDELY DISTRIBUTED THROUGHOUT EUROPE AND BECAME, IN EFFECT, A STANDARD MIXED-TRAFFIC TYPE FOR MUCH OF THE CONTINENT. A TOTAL OF SOME 3,850 LOCOMOTIVES HAD BEEN BUILT BY THE 1930S. EXAMPLES SURVIVED INTO THE 1970S AND SOME OF THESE ENDED THEIR DAYS AROUND GERMANY'S BLACK FOREST REGION.

LEFT: A China Railways industrial SY Mikado at Anshan Iron and Steel Works. Though an industrial design, the SYs are descended from the typical American Light Mikados used for mixed-traffic work on many of the American roads during the second decade of the 20th century. The type's utility in operating around steel works and collieries is an interesting cross-fertilization of two traditionally different orders.

ABOVE: One of Nigel Gresley's superb V2 mixed-traffic 2-6-2s on test. One of the most successful designs, these three-cylinder engines were referred to as the "engines which won the war" on account of the prestigious feats of haulage they performed during the crucial years between 1939 and 1945.

OPPOSITE BELOW LEFT: The driver adds a round of oil to the moving parts of a Robert Stephenson and Hawthorn 4-6-0 in Brazil.

OPPOSITE BELOW RIGHT: The Walschaerts valve gear of a British Railways Standard 4 mixed-traffic 4-6-0.

Above: The 4-6-0 was a principal mixed-traffic concept and one of the most brilliant designs was Stanier's Black 5 of the LMS of which 842 were built. Although they were not suitable for heavy mineral haulage, the Black 5s, with their 1.8 m (6 ft) diameter driving wheels, performed express passenger and freight working to perfection. Their performance was often indistinguishable from designs specifically produced for passenger work.

Preservation

Saved from the Cutter's Torch

Britain and America lead the world in railway preservation with exhibits dating across one-and-a-half centuries of development. "Locomotion" of 1825 was the first main-line locomotive and ranks alongside the "Flying Scotsman" a century later as a milestone in the development of steam traction. What a fine contrast they make with the classic outside cylinder 4-4-0 "William Crooks", one of the definitive early American types featuring cow catcher, brass bell and huge spark-arresting chimney.

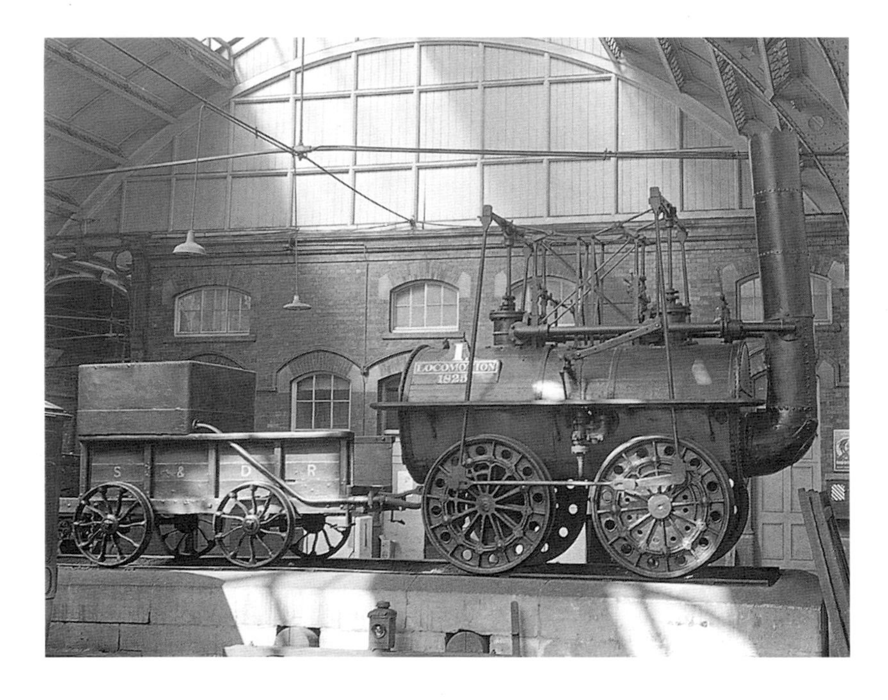

Left: George Stephenson's "Locomotion" on display at Darlington Bank Top station. This engine hauled the first train on the Stockton to Darlington Railway' in 1825. It was sold for scrap in 1850 but was subsequently used as a pumping engine before being restored.

Above: "William Crooks" – a typical American West, mid-19th century 4-4-0, built at Paterson in New Jersey – is a prime exhibit at Duluth Railway Museum.

Railway preservation is dependent on volunteer enthusiasts. Without them hardly anything would remain of the steam age except perhaps a few official collections, probably static displays in museums. Fortunately, millions of people revered the sight and sound of working steam, especially in Western countries. When it began to disappear, a vigorous rearguard action was fought to save as much as possible. Locomotives and rolling stock were retrieved from scrapyards at an incredible rate. Now in Britain alone there are over one hundred railway companies operating steam trains, which has created a vast new area of tourism and brought the magic of steam to new generations. Over the years, many branch lines and secondary routes which were closed down in the early 1960s became available to enthusiasts. Preservationists managed to acquire many of these lines and have slowly brought them back to life as living, operating railways which, in some cases, almost mirror the line in its heyday over a hundred years ago.

Centre: Works plate of a Bombay, Baroda and Central India Railways locomotive at India's National Railway Museum in New Delhi.

Left: The "Flying Scotsman" is probably the world's most famous steam locomotive. It was one of Gresley's prestigious A3 Pacifics, 78 engines in all, which worked the crack expresses of the LNER.

Opposite: The former Midland Railway roundhouse at Barrow Hill near Chesterfield featuring standard Midland 0-6-0 shunting tanks along with one of the ubiquitous 4F 0-6-0s and an L&Y Dock Tank.

47630
44482
41708
740

PRESERVATION

RESTORING THE ORIGINALS

CANADA, AUSTRALIA AND SOUTH AFRICA ALL HAVE THRIVING STEAM PRESERVATION ORGANIZATIONS, COVERING A WIDE RANGE OF ENGINES. IN THE THIRD WORLD, COUNTRIES ARE MORE INTERESTED IN LOOKING FORWARD, AND THERE IS NOT THE WHEREWITHAL TO SPEND ON PRESERVING THE PAST. A FEW, SUCH AS SIERRA LEONE, HAVE ABANDONED RAILWAYS ALTOGETHER; OTHERS, SUCH AS GHANA, HAVE DESTROYED THEIR STEAM AGE VIRTUALLY WITHOUT TRACE. PARADOXICALLY, ERITREA, ONE OF THE WORLD'S POOREST NATIONS, IS ACTIVELY REBUILDING ITS ABANDONED RAILWAY SYSTEM, PRIMARILY WITH TOURISM IN MIND.

Brilliant though the work of the preservationists is, schemes are invariably carried out piecemeal and never as part of any co-ordinated national plan. Also, railway preservation is a phenomenon of recent decades and famous types which vanished in the 1920s, 1930s and 1940s are thus excluded.

Some American railways, such as the Southern Pacific and Union Pacific, had an enlightened policy towards preserving their heritage, but others, specifically the Eerie Road and, tragically, the New York Central, preserved nothing. Despite America having some sixty Class 1 railroads at the end of the steam age, there was no co-ordination of preservation, no corporate policy and no central funding. This meant that a random selection of locomotives was preserved, many of which ended up in private sidings or public parks. Once the novelty wore off, some were left to go derelict, others were cut up and some continue to be well maintained.

In all, North America has some 1,500 steam locomotives, about 150 of which are in running order – far fewer than in Britain but, *per capita*, Britain's interest in railways exceeds the interest anywhere else in the world, and connections with the railway past are strong and unique. In North America, upwards of 75 locomotives regularly run each summer on preserved lines and rail tours. America has many fine railway museums, some state-run, some private, and some operated by foundations; there are over 200 sites at which steam locomotives can be seen nationwide.

Opposite left: Another classic American 4-4-0 in the form of Canadian Pacific Railway's No. 136, pictured heading a mixed train through Berkeley, Ontario, on 14 October 1973.

Opposite right: The magic of steam in South Africa is caught in this study of double-headed Class 24 2-8-4 Berkshires departing from Miller with the Indian Ocean Limited special on 28 April 1991.

Above: In contrast, a pair of Canadian Pacific Railway's GS Class 4-6-2s. This pair of late era Pacifics are seen heading a Steamtown excursion train near Brockways Mills, Vermont.

Right: A brace of Australian C30 Class 4-6-0s in Woodhouslee to mark the last train to Crookwell in August 1989. The leading engine is preserved at Thirlmere Railway Museum and the train engine in Canberra. Built in Manchester, England, by Beyer Peacock, these engines reflect that maker's traditional concern with beauty of appearance.

Preservation

Building Replicas

One of the most exciting aspects of contemporary railway preservation is the building of replica locomotives to compensate for historic types which no longer exist. Although this practice is not new, it is playing an increasingly big part in the preservation ethos. The building of a British LNER A1 Class Pacific at Darlington is by far the most advanced project to date, and has captured the imagination of enthusiasts the world over. When completed, the new A1 will act as an important catalyst for new schemes which will bring both variety and vitality to the preservation movement in general.

Preserved railways give pleasure to millions throughout the world, and have become an international tourist attraction in their own right. They introduce the magic of steam to new generations who might otherwise regard it as a dead phase of a distant and irrelevant history. Equally importantly, they attract attention to the railway itself and this brings obvious benefits to our modern railway industry, which must win hearts and minds if rail is to achieve its rightful place as the world's principal form of land transportation.

OPPOSITE LEFT: A working replica of the world's first steam locomotive, Richard Trevithick's engine built for the Coalbrookdale Ironworks in 1803. In the autumn of 1998, it was in action at Blists Hill which forms part of the Ironbridge Gorge Industrial Heritage site.

OPPOSITE RIGHT: A Fairlie 0-4-4-0 articulated in action on the 0.6 m (1 ft 11½ in) gauge Ffestiniog Railway in North Wales. This system was originally a slate-carrying line. Fortunately, the final decline of slate traffic in the mid-20th century coincided with the rise in tourism, which enabled the railway to continue as a passenger-carrying line from 1955. Along with the Talyllyn Railway further south, the Ffestiniog was one of the world's pioneering railway preservation schemes.

ABOVE: This working replica of "Tom Thumb" is one of America's most famous preserved engines. "Tom Thumb" was designed in America in 1830 and the replica is seen here on its original track at the Ellicott City terminus of the Baltimore and Ohio Railroad.

RIGHT: The 0.9 m (3 ft) gauge Isle of Man steam railway is a major attraction on the island and one of the most widely promoted preserved railways in the world. The line operates with some of the original locomotives in the form of 2-4-0Ts, which came from Beyer Peacock's works in Gorton, Manchester. The example shown here is No. 4 "Loch", built in 1874.

Index and Picture Credits

Picture Credits:
All pictures copyright Milepost 92½ except those listed below:
Martin Pemble/Milepost 92½ pages 2, 44tl, 45, 53, 60br; Lady Gretton page 12;
Ted Smart page 28; Arthur Mace/Milepost 92½ page 34t, 56r, rear jacket centre, end papers;
Ron Ziel pages 36l, 37br, 43bl, 54tr, 60l, 61t; John R. Jones page 48t; Brian Soloman/Milepost 92½ pages 58tr, 63tl;
Leon Oberg/Milepost 92½ page 61br; Brian Burchell/Milepost 92½ front jacket; Fred Kerr/Milepost 92½ page 59